Where, oh where, does the yellow corn grow?

by Nick and April Patterson

Illustrated by Karen Smith
Edited by Claudia Johnson

www.acresofgracefarms.com

Where, oh where,
does the yellow corn grow?
It starts with seeds
planted all in a row.

Nestled deep
in the soil
is where it all starts.
Just add rain,
sunshine, and love
from the
farmer's heart.

The seeds send down a root
and come up as a sprout.
Soon after that,
the leaves emerge out.

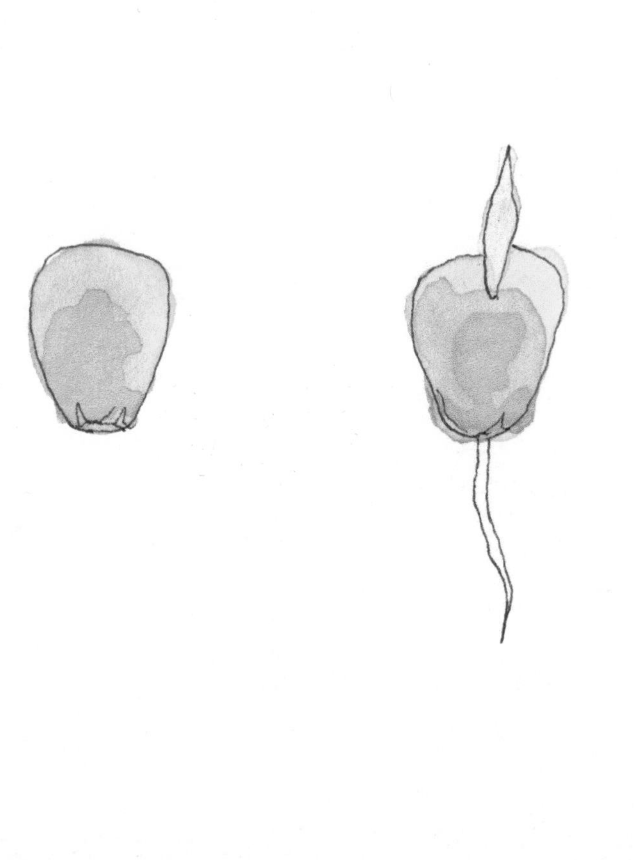

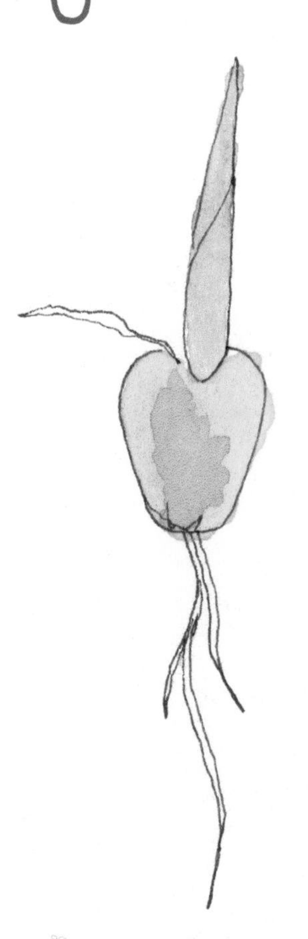

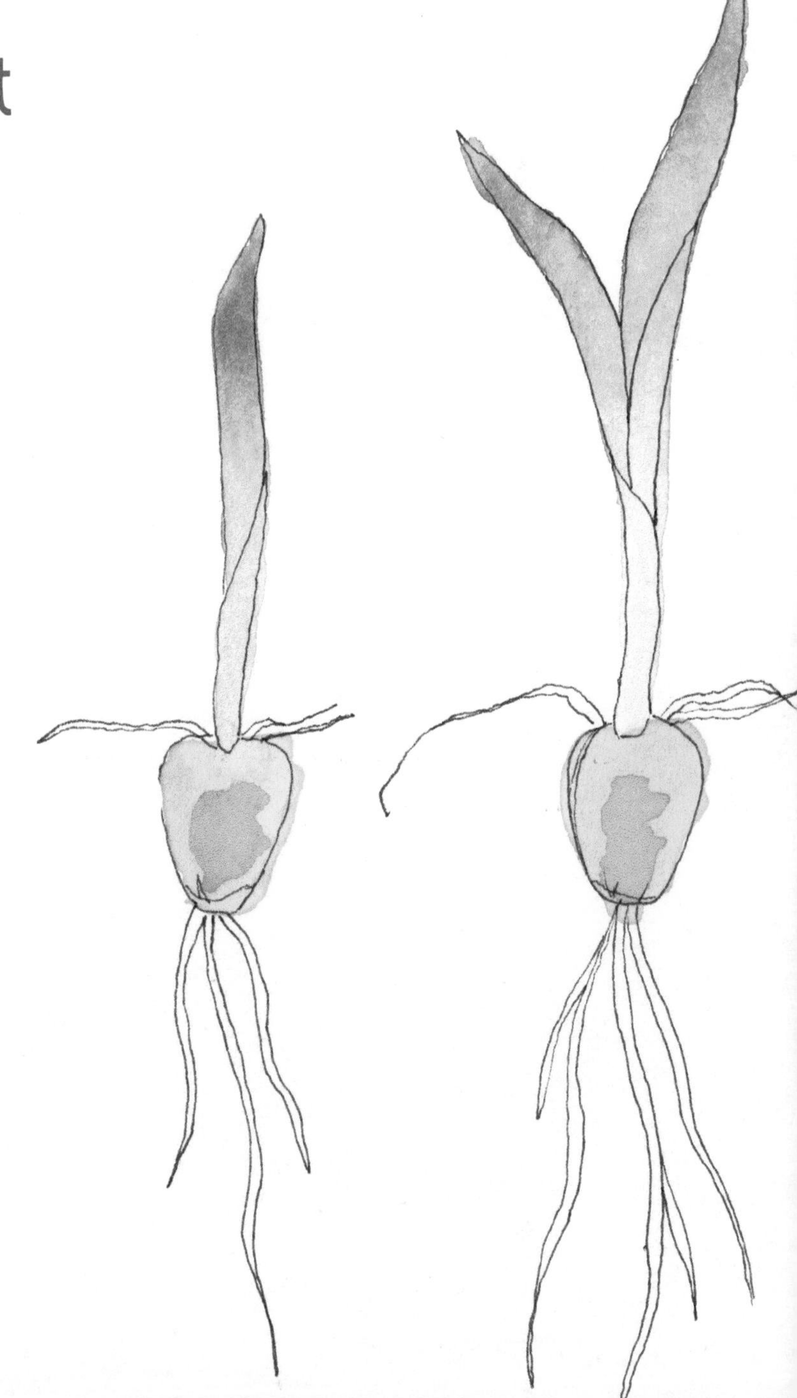

Now the farmer's work truly begins,
spraying and feeding all his new plant friends.

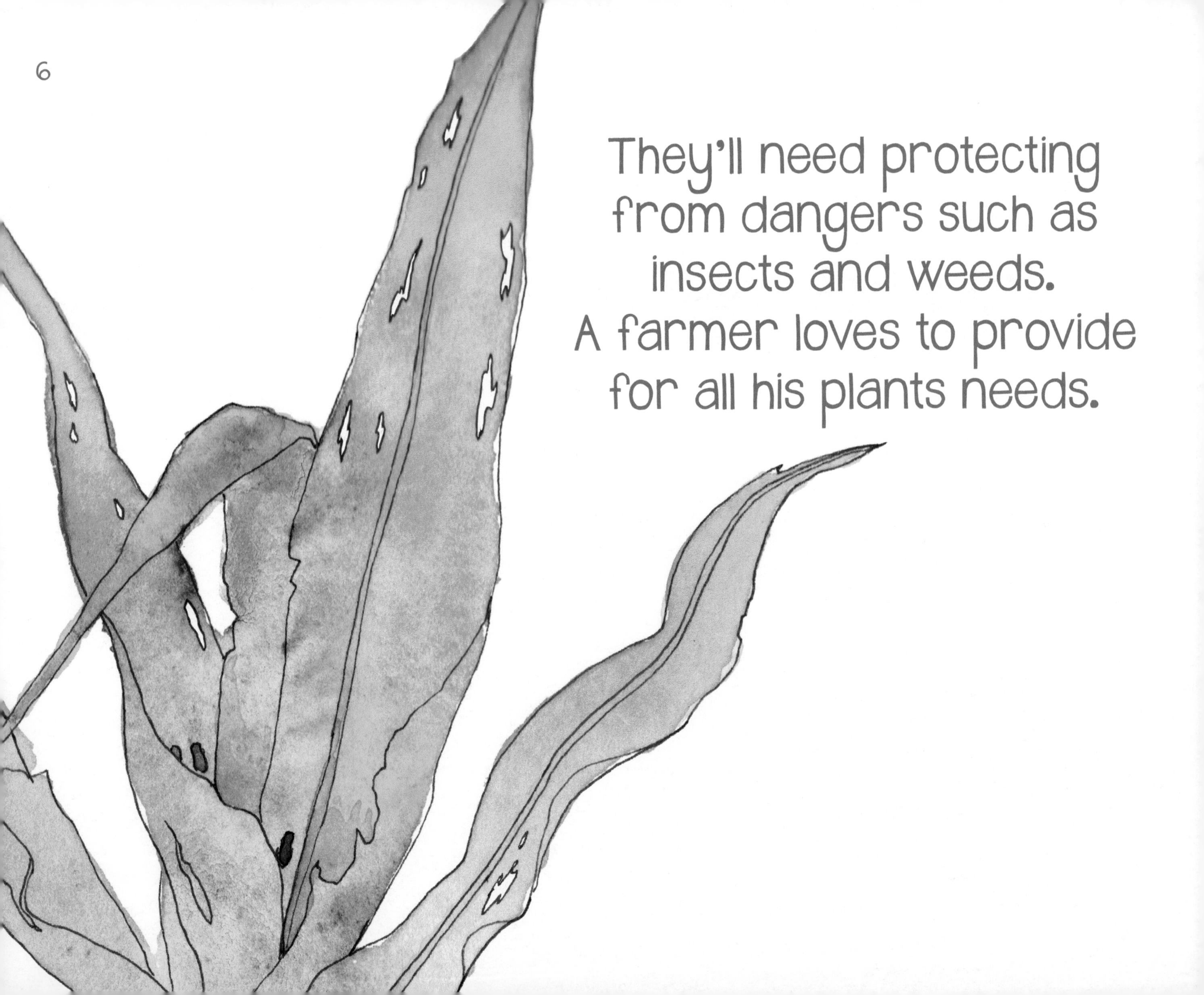

They'll need protecting
from dangers such as
insects and weeds.
A farmer loves to provide
for all his plants needs.

The plant will
grow up to
twelve feet tall,
making ears
full of kernels
right before fall.

As the seasons change
so do the plants.
Soon the leaves and ears
will quickly dry out.

When the winds of fall
begin to blow,
the farmer on his combine
drives down all the rows.

This gigantic machine
is as big as a bus!
It takes the kernels off the cob
without any fuss.

From the combine to a grain cart
the kernels now go -
next into large trucks
for transport down the road.

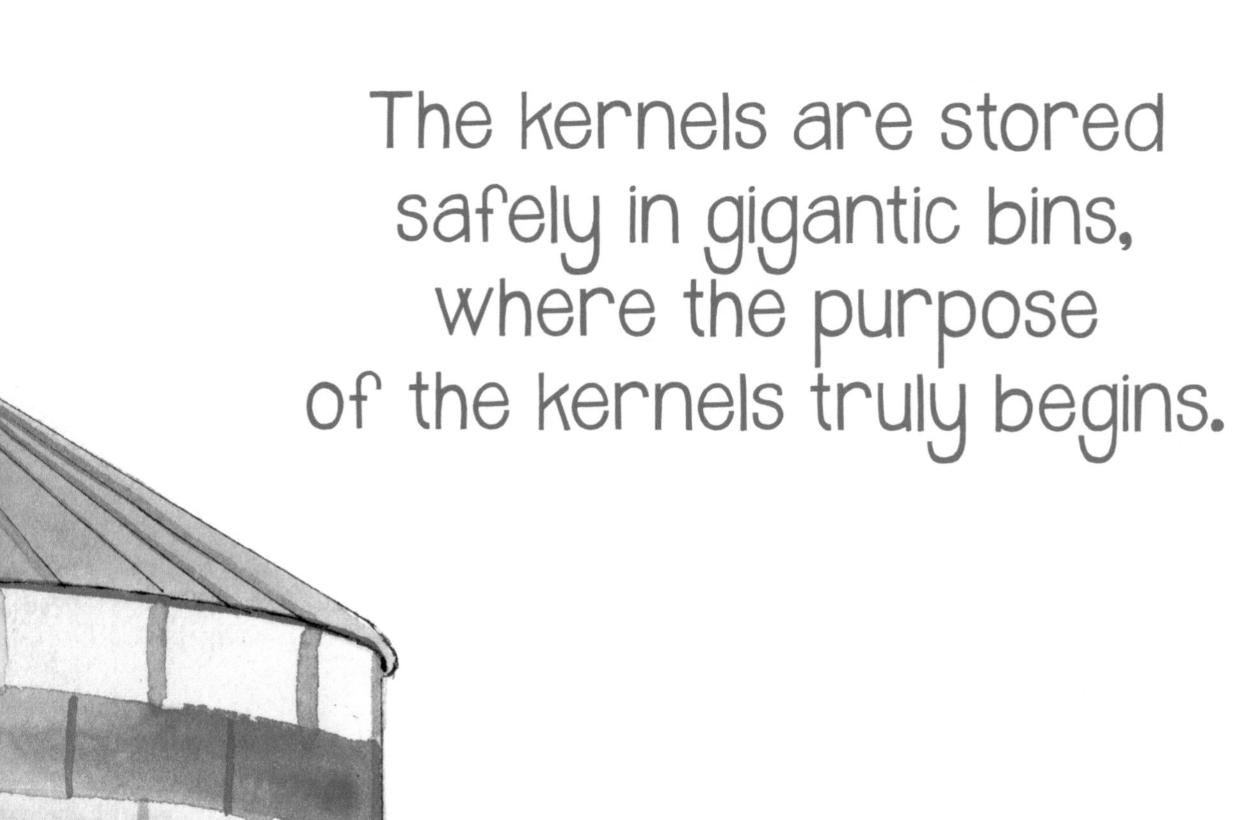

The kernels are stored
safely in gigantic bins,
where the purpose
of the kernels truly begins.

The corn is so happy
to help keep us healthy.
It provides yummy nutrition
for all of our bellies.

Cornbreads and muffins and even cornflakes
are just a few things the corn deliciously makes.

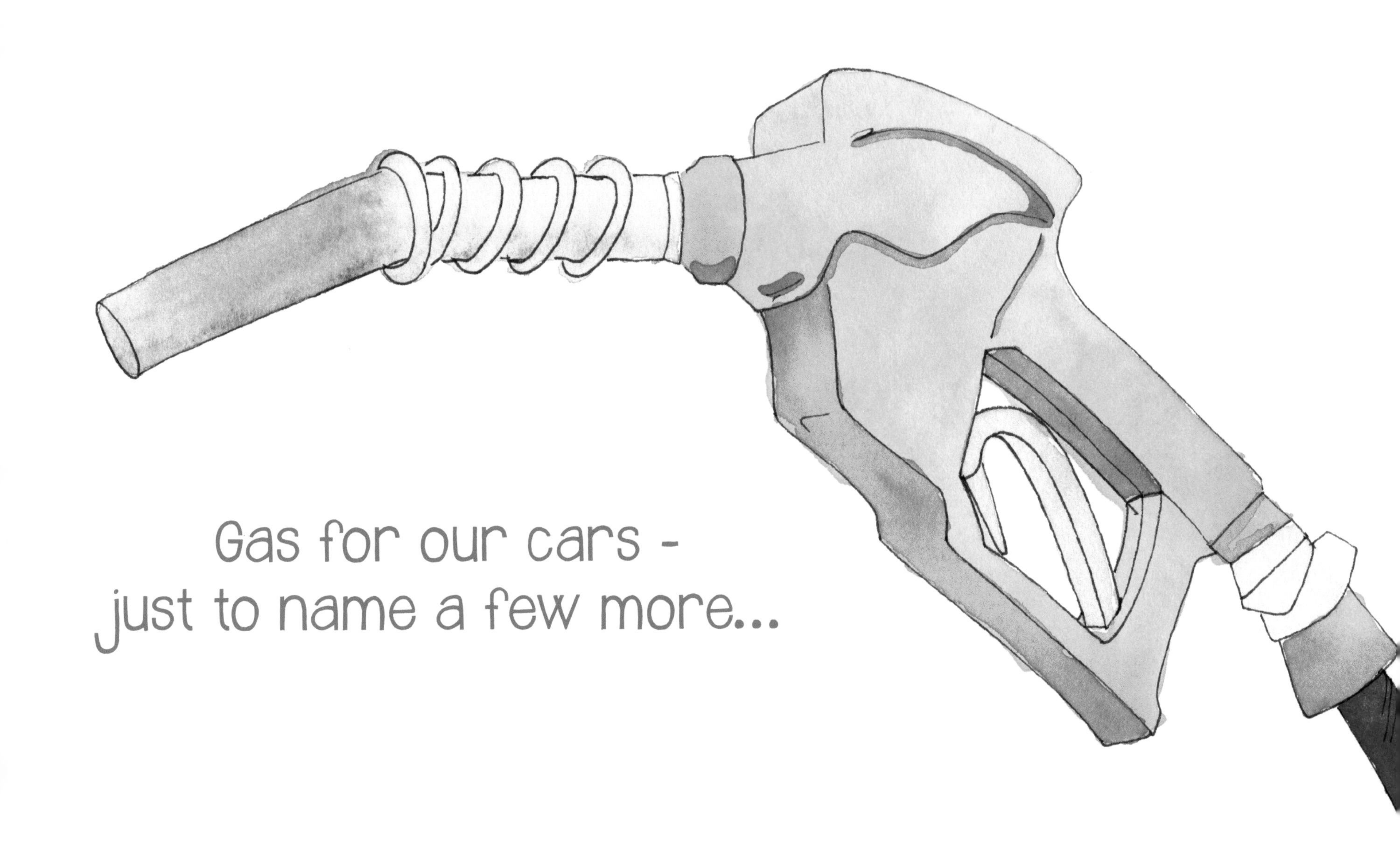

Gas for our cars -
just to name a few more...

food for our cattle...

and chickens galore.

Now you know how the yellow corn grows.
It's part of your life wherever you go.

Dedication

This book is dedicated to our parents,
Doug and Janie Smith and Randy and Cindy Patterson,
for instilling in us a love for the earth, admiration
for farmers, and a belief in the Creator and Giver of Life.

"He who tills his land will have plenty of food."

~ Proverbs 12:11

About This Book

The purpose of this book is to instill in children the importance of agriculture in their lives.
It is our desire that each young person grows up with an appreciation of the land, for the farmers that cultivate it, and for the myriad of valuable products it produces.

April and Nick Patterson

About the Authors

April Smith Patterson grew up in rural Clay County, Tennessee, with love for the beauty of the land and gratitude for its bountiful resources. When April met veterinarian Nick Patterson, she found a kindred soul. Since their marriage, the two have continued to develop their farms, called Acres of Grace, raising livestock, crops, chickens, and more.

In addition, April is a partner with her mother and brother in a number of companies founded by her late father, Doug Smith, wih her primary responsibilities being oversight of Honest Abe Log Homes. Nick serves as president of another Smith Family company, Barky Beaver Mulch & Soil Mix, which produces mulch and soil products for commercial growers and retail sales. Nick, who was raised on a farm in rural North Alabama, is the third generation to work his family's land alongside his father and Uncle Den. He feels blessed to have the opportunity to serve as chairman of the Clay County Young Farmers & Ranchers and be a part of the agricultural community.

The Pattersons also own The Mill Storehouse, a retail store in Algood, Tennessee, and online, offering clothing, home and garden decor, farm produce, original items and much more.

Thank You

We wish to express our gratitude to Tennessee Foundation for Agriculture in the Classroom.
This book was inspired by the dedication exhibited
by the teachers, administrators and parents who work together
to educate our children and youth about the
importance of farming and the value of farmers.

April & Nick Patterson

This book is made possible through the commitment of these Tennessee companies.

themillstorehouse.com

acresofgracefarms.com

FOR REPRINTS OF THIS BOOK, CONTACT NICK AND APRIL PATTERSON

Phone: 1-800-737-3646

Mailing Address: 9980 Clay County Hwy., Moss, TN 38575

Email: April@acresofgracefarms.com